Leif Catches the Wind

A Mechanical Engineering Story

Written by the Engineering is Elementary Team
Illustrated by Jeannette Martin

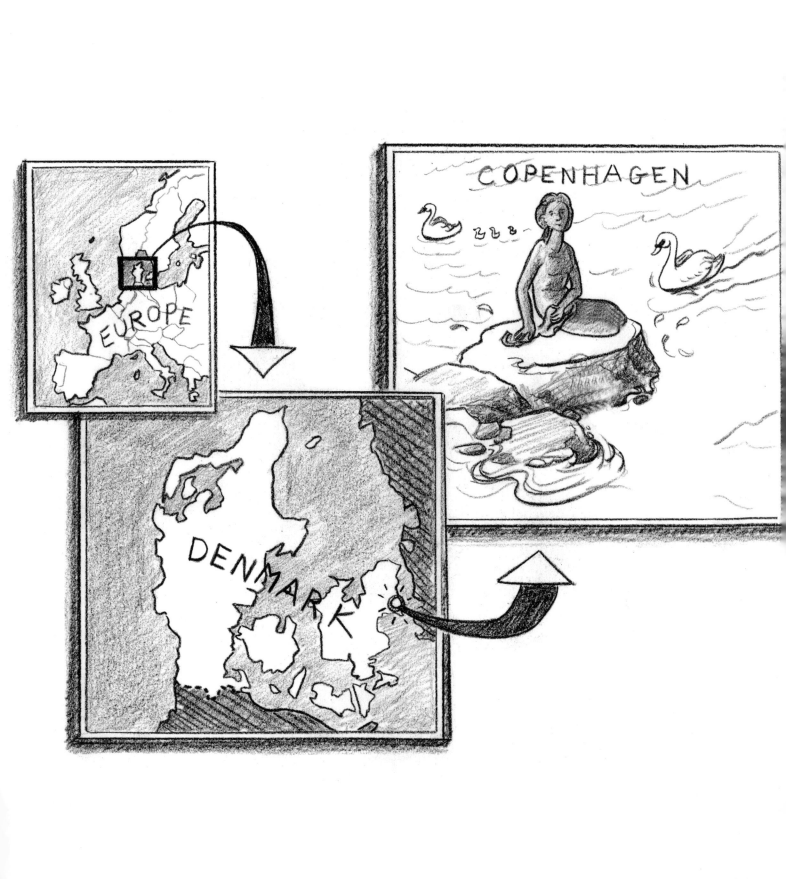

Chapter One | Lonely Day

Subject: Hi!
From: Weathergirl@denmarkemail.com
To: Fishkid@denmarkemail.com

Hi Leif,

Here I am in my new house. ☺ and ☹. I feel like I'm really far away from Copenhagen.

Mom and Dad are still unpacking. They let me set up my weather station first thing! Then we set up the computer so I could email you. Remember, you promised to send me *vejr* information—weather information—every day. I'll send you a report from here, too. Then I'll get my weather website started!

I miss you lots and lots!
Dana

Just as Leif was about to close his email, a new message popped up on his screen. It was Dana again.

> **Subject: One more thing**
> **From: Weathergirl@denmarkemail.com**
> **To: Fishkid@denmarkemail.com**
>
> I can't wait for your visit next weekend!
> Dana
> P.S. Tell Frederik and Joachim "hi" for me.
> P.P.S. Did you know this place has a fish pond? Now I know you'll come visit.

Leif touched his freckled nose to the cool surface of the glass fishtank next to the desk. "Hey, Frederik and Joachim—Dana says hi." Of all his pets, these two *fisk* were his favorites, because Dana had given them to him on his birthday last year. He had named them after the two princes in the Danish royal family.

Leif's mom and his auntie, Dana's mom, were twin sisters and best friends. Ever since Leif and Dana were babies, they had been best friends, too. Sometimes Leif thought he and Dana were more like twins than cousins. Just this week Dana had moved away from Copenhagen to a new

home in Aalberg. He already missed her a lot, but watching Frederik and Joachim made him feel a little better.

Leif turned around and looked at his other fish. There were platys, catfish, goldfish, guppies, tetras, and tiger fish—forty fish in all—living in five tanks. They all seemed healthy, swishing around with glistening scales.

Leif drifted around the room, trying to think of something to do. *I might as well check the weather readings and report to Dana,* he thought. She loved weather watching and forecasting as much as Leif loved raising his pet *fisk*. When she heard she would be moving away, Dana managed to turn her fascination with *vejr* into a fun way to keep in touch.

"Leif!" she had announced. "Let's make a website comparing the weather in Copenhagen to the weather in Aalberg. Maybe we can even collect and post weather reports from other kids across Scandinavia!" She had insisted that they use their special handshake to seal his promise to report the weather daily.

Chapter Two | # An Email S.O.S.

Leif looked around his room. Before Dana had left, she had given him posters, equipment, and a weather notebook. "It's your new weather station!" she'd exclaimed, sweeping some magazines off his desk to make room for the instruments. She had made many of the instruments herself.

He checked each of the weather station components and recorded the details in the notebook. Date and time. Temperature. Cloud observations. Rain report. Wind speed. Leif picked up his new anemometer—the instrument that Dana had helped him build to measure wind speed. "When the wind blows into these little cups," she had explained, "it makes the whole thing spin—quickly

when the wind is blowing fast, slower when it's not. Because we know that your anemometer and my anemometer have the same components, we can compare our wind speed results!" *Leave it to Dana to think of everything*, Leif thought.

After checking the weather, he carefully recorded all the information in his weather notebook, then turned on the computer to send his report to Dana.

> **Subject: RE: Hi!**
> **From: Fishkid@denmarkemail.com**
> **To: Weathergirl@denmarkemail.com**
>
> Dear Dana,
>
>   It's NO fun without you. Here's my *vejr* report for today:
>
> Time: 12:53 PM
>
> Temperature: 8 degrees Celsius
>
> Clouds: none
>
> Rain: none
>
> Wind speed: 20 Kph (Strong wind)
>
> I'm bored,
>
> Leif
>
> P.S. Joachim and Frederik say hi.
>
> P.P.S. 10 days until I visit.
>
> P.P.P.S. What kind of fish are in your pond?

He clicked SEND. Before Leif could log off, the email inbox chimed. Dana was online!

> **Subject: RE: Hi!**
> **From: Weathergirl@denmarkemail.com**
> **To: Fishkid@denmarkemail.com**

Leif,

    I'm so glad you're online! I have a new weather assignment for you. Go to the harbor and get me wind data, okay? Not just anemometer readings, but also Beaufort scale. (Remember when I showed you how to use the B. scale? You can use it to figure out wind speed based on what's happening around you.) Report back. I think a front might be on its way to you.

    One other thing. The *fisk* we have are big goldfish. But they don't look so good. Their gills are pumping hard! Can you help? What do I do?

Miss you,
Dana

P.S. And I forgot: Go fly a kite! (tee hee)
P.P.S. I mean it. Tell me how it flies. Good for my website, okay?

Chapter Three | Harboring Ideas

Leif's hair streamed back from his face as he raced against the wind. He coasted past the "Little Mermaid" statue at the harbor park and past the wind turbines out in the harbor that produced electricity for the people of Copenhagen. He and Dana used to love coming to the harbor to watch the blades of the turbines spin, like beautiful pinwheels rising out of the water.

The harbor was where Dana had first taught Leif to pay attention to weather. She always pointed out the big, puffy cumulus clouds that looked like fish, mermaids, and other creatures. Sometimes, because of the forecasting she did, she would call ahead of time

and tell him to bring his kite, even when there was no breeze. By the time they had finished their picnic lunch, a new weather front would have moved in and there would be enough wind for flying kites.

Leif hopped off his bike, gently tossing his backpack onto the ground. As he unrolled his kite and string, he looked around. Whitecaps tipped the waves in the harbor. The leaves were tossed by the breeze, and he noticed that smoke from a distant chimney was streaming out horizontally. From his practice with Dana, he had the Beaufort scale memorized. Right now, it was about 4 on the scale, which meant that the wind was blowing 20–28 kilometers per hour—about as fast as he'd been pedaling to get here.

Catching the wind, his bright orange, fish-shaped kite tugged at the string in Leif's hand. The air seemed to scoop it away, and it took off. Watching the kite "swim" reminded him of Dana's new fish. She had said their gills were pumping hard. What could be the problem?

He remembered the time he had almost lost a whole tank of *fisk*, after the power went out in his house for days. After the fishtank pump had stopped working, the fish had flapped their gills furiously.

That's when he had learned that fish need oxygen, too, just like people. But fish take in oxygen from the water instead of from the air. The pump in the fishtank helped mix air into the water, so the fish could get more oxygen than from water alone. Leif stared towards the wind turbine blades spinning in the breeze. As a motorboat passed by, he noticed its propeller blades spinning just like the turbine blades—only they were stirring up the water behind the boat. Leif suddenly smiled. He had an idea to help Dana's fish.

Harboring Ideas

Chapter Four | An Engineering Talk

Leif knew he wanted to make a paddle to stir up the water in Dana's pond and give the fish more oxygen. The question was, how could he power a paddle? Leif knew just the right person to ask: his mom. She was an expert at generating electricity from the wind.

Leif's mom was a mechanical engineer who worked on wind energy projects. Engineers are people who combine creativity with their knowledge of math and science to solve problems. Leif's mom helped figure out how the different parts of wind turbines and other machines should be designed so they fit and worked well together. Leif's mom liked the

An Engineering Talk

challenges of her job—deciding just the right way to put the machines together—and she liked working on teams with lots of other engineers.

Leif figured that Dana's goldfish had a problem worth solving. When he got home, he found his mother in the kitchen.

"Mom, how could I generate energy with a windmill?" he asked.

His mom turned to him from her cutting board, her hair falling over one eye. "What is it that you need the energy for?"

"I want to power a paddle. A water paddle."

"Okay," she said, "so what you really mean is that you want to change wind energy into energy that will power—or move—your paddle." Leif's mom was always so choosy about the words people used. "You know that there's energy in the air. The hot and cold spots of air cause currents, or motion." She whooshed one hand past the other.

"You mean the wind?" Leif asked.

"Yes, the wind. So the energy of the wind can be used to do something useful."

"Right. Like turn a windmill."

"A windmill or a turbine," Mom said.

"What's the difference?" Leif asked.

"A wind turbine changes the air movement into a spinning motion that is used to generate electricity."

Leif shrugged and picked a carrot stick from Mom's pile. "Doesn't a windmill spin?"

"Yes, but a windmill doesn't generate electricity. The turning motion does other things, like work a grindstone to grind flour, or move the parts of a water pump. It sounds like you want to do something similar to pumping water. You want to get the water moving, anyway."

Leif crunched the carrot. "So how could I make the windmill do that?"

His mom put the knife down and folded her arms. "That depends on a lot of things. How big will the paddle

be? What materials are you using to make the windmill and paddle? How will the windmill and the paddle be connected?"

Sometimes his mom could be too picky. "I don't know exactly. What do you think I should do?"

She chuckled and tapped the tip of Leif's nose with her finger. "I think you should try to figure it out. I can help you. How about if you design a plan for your windmill and build a model?"

An hour later, Leif's mother was leaning over him, talking a mile a minute.

"Oh, that's good, Leif. You got the blades to spin like a pinwheel!"

"Or an anemometer!"

She looked surprised. "Oh, yes, you must've learned that because of the *vejr* station you and Dana set up. You're right. Pinwheels, anemometers, windmills, and turbines all have parts that catch the wind. You've got the basics down."

She turned the model around to view the back of it. "To get this to work outside you'll have to think about some sturdy materials to use. Then you have to experiment with it until you find something that works. It's a big project, Leif."

Leif felt like a fish swimming in a swift river of possibilities.

"I know it is, Mom. But I bet I know someone who can help me," Leif said, trotting off to turn on his computer. He needed to talk to Dana so the two of them could brainstorm together. *Two heads are better than one*, he thought.

Chapter Five: Working Together

Internet Message
From: Weathergirl
To: Fishkid

Weathergirl: Hey! Did you get my weather data?

Fishkid: Yeah. I went over to the harbor to collect it. And while I was there, I had a great idea. We're going to design a windmill that turns a water paddle in your pond.

Weathergirl: A what?

Fishkid: A water paddle. It can stir up the water. You know how the propellers on motorboats churn up the water in the harbor? The paddle will churn up the pond water and mix air into it, like making whipped cream with a whisk. And the windmill will power the paddle.

Weathergirl: I don't get it. How is this windmill and paddle contraption going to help the fish?

Fishkid: The fish are sick because they're not getting oxygen. That's why Joachim and Frederik have a pump in their fishtank—to pump air into the water. If we use a paddle to churn air into the water in your pond, the fish will get more oxygen!

Weathergirl: Great idea! So now what?

Fishkid: I'm not sure. Mom keeps telling me to use the engineering design process.

Weathergirl: Huh?

Fishkid: It's ASK, IMAGINE, PLAN, CREATE, IMPROVE. She's always trying to get me to do that, even when we're just building toy sailboats.

Weathergirl: Maybe we should try it her way this time. The fish are depending on us to save them!

Fishkid: Okay. So let's both ASK and IMAGINE and we'll talk tomorrow morning.

Weathergirl: Okay! What should we ask about?

Fishkid: What kinds of materials will work outdoors, and how to get a nice fast spin on the windmill, and then maybe what might work for the paddle part?

Weathergirl: Got it. Miss you.

Fishkid: I miss you too.

Moonlight poured through the window as Leif tossed and turned. His dreams had been of fish gasping for air. He couldn't let the night go by without working to rescue them. Leif slipped out of bed and crept over to his craft box.

He picked up his half-made windmill from the afternoon and tried spinning the blades with his hands. It worked, but the wooden spindle was rubbing against the ragged hole cut from the cardboard carton. *What was it*

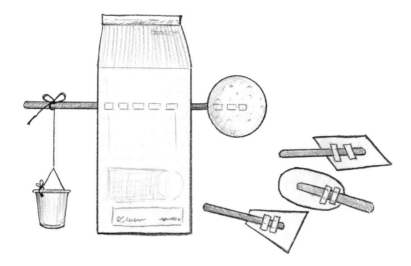

Mom had said about that rubbing? Too much friction. It kept the blades from moving freely. Leif rummaged through the craft box for ideas. What could he put around the spindle to make it more slippery?

He picked up some masking tape, felt its outer surface, and placed it aside. How about the shiny fabric? He tried that, but its fibers caught on the cardboard. At last he found something that worked.

Now that the friction was low, Leif could see other problems. The whole machine was wobbly. He wasn't sure the pinwheel-shaped blades would catch the wind well. *Imagine, imagine, imagine*, he thought to himself, but no new ideas came to him. *Maybe I should sleep on it*, he thought.

The next morning Leif woke up and headed to the computer.

> **Internet Message**
> **From: Weathergirl**
> **To: Fishkid**
>
> **Weathergirl:** My windmill is wobbly. And the blades are only paper. That won't hold up outdoors. What kinds of blades did you use? Any idea what kind of wind speed we need?
>
> **Fishkid:** My blades were paper, too. I folded them like pinwheels. I don't know the wind speed we need, but at least you can get an idea of what kind of wind speed we can count on, right? With all your weather records, can you take a look? We just need to make the windmill so it can move a paddle in the water.
>
> **Weathergirl:** Can you send me your plan? The fish are still looking sick. We've got to work fast. Leif, do you think we can really do this?
>
> **Fishkid:** I'll write up my plan and send it to you later today. I KNOW we can do this!

Just before lunch, Leif clicked SEND and smiled. He had managed to draw up a plan, even if it wasn't perfect. So along with his weather report, his plan was on its way to Dana. Now each of them would build a model and test it. The results would help them tackle some of the remaining problems.

An hour later, satisfaction was washing over him. The windmill blades went around so fast that they were a blur. By combining his ideas with Dana's ideas, he had designed a stable, sturdy, and reliable windmill. And what's more, he had gotten the windmill to do some work—it could lift an action figure off the floor!

Now he and Dana could work on improving the design. First, they would have to find a material for the blades that worked as well as the paper, but could withstand the outdoors. And then they could make a plan for what his mom would call a prototype. The prototype would be a full-size plan for a windmill that would turn a paddle to churn the water, instead of lift a toy. He and Dana still had lots of emailing and designing to do, but he was encouraged that they could finish in time to save the *fisk*.

Working Together

Chapter Six | Saving the Fish

A few days later, the wind was cool against Leif's face as he watched healthy *fisk* shoot forward in Dana's pond. Their tails swished. Their scales caught bright blazes of sunlight.

"Hey, fishies, meet your superhero! Leif, the junior mechanical engineer!" Dana called, tossing some crumbs into the pond.

Leif laughed. "You're the one who figured out how to make the windmill stand up straight!"

"But you came up with the idea of a windmill-powered paddle in the first place, and you figured out a basic plan. All I did was help improve the prototype."

"I guess we're both junior engineers. Mom did say that engineers usually work in teams," Leif said. "I'm just glad we could still do stuff together even though you moved away."

Across the clearing, Leif's mother and auntie were laughing. "Just look at those two!" he heard his auntie say. "Do they remind you of anyone?"

His mom's smile was warm as she caught his eye and waved him over. "Cousins or not, they are twins, just like us. At least they put things together. Remember how we always used to take things apart?"

The two women laughed. Leif had a sudden thought. His mom and auntie had lived apart for many years—sometimes close by, sometimes far away—but they were still good friends and still had fun together. A big load of worry slipped off his heart. He called over his shoulder to Dana—"Race you to lunch!"—and took off with the wind.

Try It!

Design a Windmill

Windmills can use energy to do all sorts of things, such as grinding grain, pumping water, and lifting a weight. Using what you learned from Leif and Dana, your goal is to design a milk carton windmill that can lift a small toy.

Materials
- ☐ Empty milk or juice carton
- ☐ A thin dowel or stick
- ☐ String
- ☐ Paper cup
- ☐ 3-inch foam ball
- ☐ Craft sticks
- ☐ Paper and index cards
- ☐ Pebbles, pennies, or other small weights

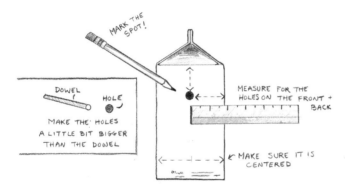

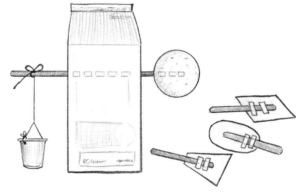

Create Your Windmill Base
Ask an adult to help you make holes in the front and back of your carton. Use a ruler to measure and mark where the holes should go. Punch the holes with a pen or nail and make them a little bit bigger than your dowel. Place the dowel through the holes. Push the foam ball onto one end of the dowel and use a piece of string to tie a cup to the other end. Put some weights inside the carton so it won't tip over.

Design Windmill Blades
How many blades should your windmill have? What shape should they be? Try designing a few different windmill blades using paper and craft sticks. Push them into the foam ball on the end of the dowel. If it's a windy day, try bringing your windmill outside. If it's not windy, you might need to use a fan. Which blades catch the wind and start to spin? Try placing some pennies or pebbles in the cup tied to the dowel. Can your windmill lift the weight?

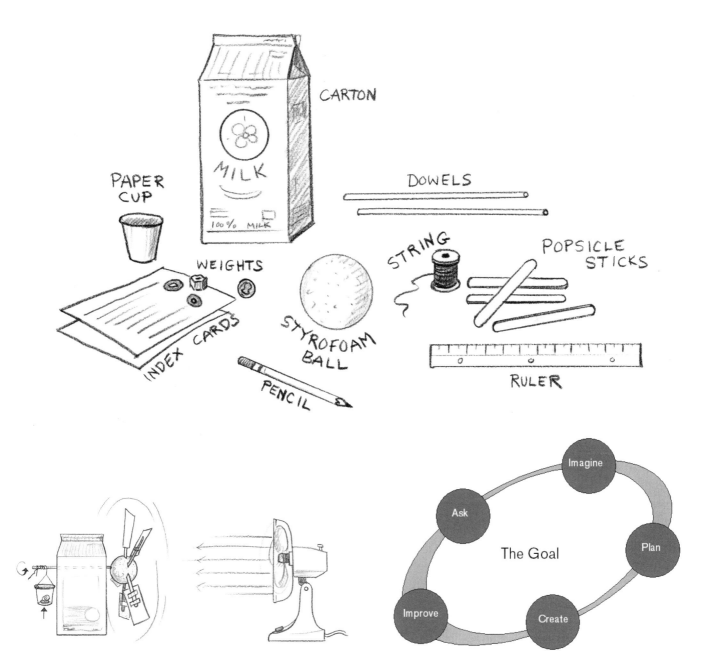

Improve Your Windmill Blades
Can you change your windmill blades so that you are able to lift even more weight? What happens when you change the blade shape? How about when you change the material that you use to make the blade?

Learn More
Go to the library or look on the internet to learn more about windmills. After you've done some research, try going through the engineering design process again.

See What Others Have Done
See what other kids have done at http://www.mos.org/eie/tryit. What did you try? You can submit your solutions and pictures to our website, and maybe we'll post your submission!

Glossary

Anemometer: An instrument used to measure the speed of the wind.

Beaufort Scale: A rating scale used to measure wind speed through observations.

Energy: The ability to do work.

Engineer: A person who uses his or her creativity and understanding of mathematics and science to design things that solve problems.

Engineering Design Process: The steps that engineers use to design something to solve a problem.

Fisk: Danish word for fish. Pronounced *fisk*.

Mechanical Engineering: The branch of engineering that deals with the design and performance of machines.

Technology: Any thing or process that people create and use to solve a problem.

Vejr: Danish word for weather. Pronounced *vair*.

Wind: Air that is moving and has energy.

Windmill: A machine that harnesses the energy of the wind to do useful work.

Wind Turbine: A kind of windmill that changes the energy of air (wind) into electrical energy (electricity).